SAFETY FOR MANAGERS

SAFETY FOR MANAGERS

A Gower Health and Safety Workbook

Graham Roberts-Phelps

Gower

© Graham Roberts-Phelps 1999

All rights reserved. No part of this publication may be reproduced, stored in a retrieval system, or transmitted in any form or by any means, electronic, mechanical, photocopying, recording or otherwise, without the permission of the publisher.

Published by
Gower Publishing Limited
Gower House
Croft Road
Aldershot
Hampshire GU11 3HR
England

Gower
Old Post Road
Brookfield
Vermont 05036
USA

Graham Roberts-Phelps has asserted his right under the Copyright, Designs and Patents Act 1988 to be identified as the author of this work.

British Library Cataloguing in Publication Data
Roberts-Phelps, Graham
    Safety for managers. – (A Gower health and safety workbook)
    1.Industrial safety – Management 2.Safety regulations
  I.Title
  363.1′1′068

ISBN 0 566 08060 5

Typeset in Times by Wearset, Boldon, Tyne and Wear and printed in Great Britain by print in black, Midsomer Norton.

Contents

CHAPTER 1   INTRODUCTION   1

  PERSONAL DETAILS   3
  HOW TO USE THIS SELF-STUDY WORKBOOK   4
  NOTES FOR TRAINERS AND MANAGERS   5
  NOTES FOR STUDENTS   6
  LEARNING DIARY   7
    LEARNING OBJECTIVES   8
    LEARNING LOG   9
    LEARNING APPLICATION   10
  HOW TO GET THE BEST RESULTS FROM THIS WORKBOOK   11
    LEARNING OBJECTIVES   13
    REASONS TO LEARN   15
    OPINION POLL   16
    OPINION POLL: REVIEW   17

CHAPTER 2   UNDERSTANDING SAFETY LEGISLATION AND PRACTICE   19

  WHAT THE LAW REQUIRES: PUTTING SAFETY FIRST   21
  HEALTH AND SAFETY LEGISLATION   23
  HEALTH AND SAFETY AT WORK ACT 1974   26
  THERE IS NO SUCH THING AS AN ACCIDENT   29
    ACCIDENT STATISTICS   31
  ACCIDENT STATISTICS   32
  WORKPLACE SAFETY AWARENESS   35
  IDENTIFYING HAZARDS AND RISKS   38
    HAZARDS AND RISKS   43

CHAPTER 3   A STEP-BY-STEP GUIDE   45

  MANAGEMENT OBLIGATIONS   47
  FOUR STEPS TO MANAGING SAFELY   48
  SET YOUR POLICY   49
    SETTING SAFETY POLICY   50
  PLAN AND SET SAFETY STANDARDS   51
  KEY POINTS ABOUT STANDARDS   53
  IMPLEMENTING STANDARDS   55
    SAFETY PLANNING AND SETTING STANDARDS   56
    ORGANIZING YOUR STAFF   57
  SAFETY AUDITING   58

	Measuring Safety	60
	Managing Safely: Checklist	61
	Self-assessment Worksheet	62
	Dos and Don'ts: Managing Safely	64
SAFETY IDEAS		**65**

CHAPTER 4   SAFETY COSTS!   67

HEALTH AND SAFETY: COST OR BENEFIT?	**69**
THE COSTS OF ACCIDENTS AND ILL-HEALTH	**70**
UNHEALTHY STATISTICS	**72**
THE MOST COMMON ACCIDENTS AND THEIR CAUSES	**73**
SAFETY IS FREE ...	**75**
THE ACCIDENT ICEBERG	**79**

CHAPTER 5   CREATING A SAFETY CULTURE   81

HOW TO MANAGE SAFETY	**83**
THE COST OF SAFETY	**85**
ATTITUDE MAKES THE DIFFERENCE!	**87**
CAUSES OF ACCIDENTS AND INJURY	**90**

CHAPTER 6   LEARNING REVIEW   93

Test Your Knowledge (1)	95
Test Your Knowledge (2)	96
Test Your Knowledge (3)	97
Case Study (1)	98
Case Study (1): Questions	99
Case Study (2)	100
Case Study (2): Questions	101
Case Study (3)	102
Case Study (4)	103
Case Study (5)	104
Case Study (5): Presentation Outline	105
Case Study (6)	106
Case Study (6): Exercise	107

**APPENDIX   SUGGESTED ANSWERS TO THE KNOWLEDGE
                        TESTS**   **109**

Chapter 1
Introduction

This first chapter acts as a record of your progress through the workbook and provides a place to summarize your notes and ideas on applying or implementing any of the points covered.

PERSONAL DETAILS

Name:	
Position:	
Location:	
Date started:	Date completed:

Chapter title	Signed	Date
1. Introduction		
2. Understanding safety legislation and practice		
3. A step-by-step guide		
4. Safety costs!		
5. Creating a safety culture		
6. Learning review		
Demonstration of safety in the workplace		
Steps taken to reduce risks and hazards		

Safety review dates	Assessed by	Date
1 month	_____	_____
2 months	_____	_____
3 months	_____	_____
6 months	_____	_____

HOW TO USE THIS SELF-STUDY WORKBOOK

Overview

This self-study workbook is designed to be either one, or a combination of, the following:

- a self-study workbook to be completed during working hours in the student's normal place of work, with a review by a trainer, manager or safety officer at a later date

- a training programme workbook that can be either fully or partly completed during a training event or events, with the uncompleted sections finished in the student's normal working hours away from the training room.

It contains six self-contained chapters which should each take about 20 minutes to complete, with the final section, 'Learning Review', taking slightly longer due to the testing and validation instruments.

It is essential that students discuss their notes and answers from all sections with a supervisor, trainer or coach.

NOTES FOR TRAINERS AND MANAGERS

For use in a training session

If you are using the workbook in a training event you might choose to send the manual to students in advance of their attendance, asking them to complete the Introduction (Chapter 1). Other exercises can then be utilized as required during the course.

For use as an open-learning or self-study tool

Make sure that you have read the workbook properly yourself and know what the answers are. Anticipate any areas where students may require further support or clarification.

Comprehension testing

Each section features one or more summary exercises to aid understanding and test retention. The final chapter, 'Learning Review', contains a set of tests, case studies and exercises that test application and knowledge. Suggested answers to these are given in the Appendix.

If sending the workbook out to trainees, it is advisable to send an accompanying note reproducing, or drawing attention to, the points contained in the section 'Notes for Students'. Also, be sure to set a time deadline for completing the workbook, perhaps setting a review date in advance.

The tests contained in the learning review can be marked and scored as a percentage if required. You might choose to set a 'pass' or 'fail' standard for completion of the workbook, with certification for all those attaining a suitable standard. Trainees who do not reach the required grade on first completion might then be further coached and have points discussed on an individual basis.

NOTES FOR STUDENTS

This self-study workbook is designed to help you better understand and apply the topic of safety for managers. It may be used either as part of a training programme, or for self-study at your normal place of work, or as a combination of both.

Here are some guidelines on how to gain the most from this workbook.

- Find 20 minutes during which you will not be disturbed.
- Read, complete and review one chapter at a time.
- Do not rush any chapter or exercise – time taken now will pay dividends later.
- Complete each written exercise as fully as you can.
- Make notes of questions or points that come to mind when reading through the sections.
- Discuss anything that you do not understand with your manager, safety officer or work colleagues.

The final chapter, 'Learning Review', is a scored test that may carry a pass or fail mark.

At regular intervals throughout the workbook there are exercises to complete and opportunities to make notes on any topics or points that you feel are particularly important or relevant to you. These are marked as:

Notes

_____

_____

_____

_____

LEARNING DIARY

Personal Learning Diary

Name: _____

Job Title: _____

Company: _____

Date: _____

> *The value of the training programme will be greatly enhanced if you complete and review the following Learning Diary before, during and after reviewing and reading the workbook.*

LEARNING OBJECTIVES

At the start or before completing the workbook, please take time to consider what you would like to learn or be able to do better as a result of the training process. Please be as specific as possible, relating points directly to the requirements of your job or work situation. If possible, please involve your manager, supervisor or team leader in agreeing these objectives.

Record these objectives below

1.

2.

3.

4.

5.

6.

> **PLEASE COMPLETE BEFORE CONTINUING**

LEARNING LOG

During the training programme there will be many useful ideas and learning points that you will want to apply in the workplace.

Key ideas/learning points	How I will apply these at work

PLEASE COMPLETE BEFORE CONTINUING

Learning Application

As you complete each chapter, please consider and identify the specific opportunities for applying the skills, knowledge, behaviours and attitudes and record these below.

Action planned, with dates	Review/comments

**Remember, it may take time and practice to achieve new results.
Review these goals periodically and discuss with your manager.**

Please complete before continuing

HOW TO GET THE BEST RESULTS FROM THIS WORKBOOK

The format of this workbook is interactive; it requires you to complete various written exercises. This aids both learning retention and comprehension and, most importantly, acts as a permanent record of completion and learning. It is therefore essential that you **complete all exercises, assignments and questions**.

In order to gain the maximum value and benefit from the time that you invest in completing this workbook, use the following suggested guidelines.

Pace yourself

You might choose to work through the whole workbook in one session or, alternatively, you might find it easier to take one chapter at a time. This is the recommended approach. If you are using this workbook as part of a live training programme, then make time to follow through any unfinished exercises or topics afterwards.

Share your own opinions and experience

We all have a different view of the world, and we all have different backgrounds and experiences. As you complete the workbook it is essential that you relate learning points directly to your own situation, beliefs and work environment.

Please complete the exercises using relevant examples that are personal and specific to you.

Keep an open mind

Some of the material you will be covering may be simple common sense, and some of it will be familiar to you. Other ideas may not be so familiar, so it pays to keep an open mind, as most learning involves some form of change. This may take the form of changing your ideas, changing an attitude, changing your perception of what is true, or changing your behaviours and the way you do things.

When we experience change, in almost anything, our first automatic reaction is resistance, but this is not usually the most useful response. Remember, safety is something we have been aware of for a long time, and consider (or fail to consider, as the case may be!) every day. As a result, we follow procedures without thinking – on auto pilot as it were. This often means that we have a number of bad habits of which we are unaware.

> *Example of change:*
>
> *Sign your name here as you would normally do:*
>
>
>
>
>
> *Now hold the pen or pencil in the **opposite** hand to that which you normally use and sign your name again:*
>
>
>
>
>
> *Apart from noting how difficult this might have been, consider also how 'strange' and uncomfortable this seemed. You could easily learn to sign your name with either hand, usually far more quickly than you might think. However the resistance to change may take longer to overcome.*

Make  Notes

Making notes not only gives you information to refer to later, perhaps while reviewing the workbook, but it also aids memory. Many people find that making notes actually helps them to remember things more accurately and for longer. So, as you come across points that are particularly useful or of particular interest, please take a couple of moments to write these down, underline them or make comments in the margin or spaces provided.

Review with others

In particular, ask questions and discuss your answers and thoughts with your colleagues and fellow managers, especially points which you are not sure of, points which you are not quite clear on, and perhaps points about which you would like to understand more.

Before you start any of the main chapters, please complete the following learning assignments.

Learning Objectives

It is often said that if you do not know where you are going, any road will get you there. To put it another way, it is difficult to hit the target you cannot see. To gain the most benefit from this workbook, it is best to have some objectives.

Overall objectives

- **Improvements.** We don't have to be ill to improve our fitness. Improvement is always possible.

- **Skills.** Learn new skills, tips and techniques.

- **Knowledge.** Gain a better understanding of safety issues.

- **Attitudes.** Change the way we think about safety issues.

- **Changes.** Change specific attitudes on behaviours and our safety procedures and practice.

- **Ideas.** Share ideas.

Here are some areas in which you can apply your overall objectives.

1. Hazards and risks

The first objective is to be able to identify safety hazards and risks. These may exist all around us and may not be readily identifiable as such – for example, the ordinary moving of boxes or small items, using a kettle or hand drill, cleaning and so on.

2. Prevention

Prevention is always better than cure, and part of this workbook will deal with knowing how to prevent accidents and injuries in the first place. Injuries are nearly always painful both in human and business terms. As well as accidents that cause us or others harm, there are many more accidents that cause damage and cost money to put right.

3. Understanding your safety responsibilities

Health and safety is everybody's responsibility, and safety is a full-time job. As you complete this workbook you will be looking at how it affects you personally and the role that you can play, not only for your own safety but also for the safety of others around you.

4. Identifying ways to make your workplace safer

A workbook like this also gives us the opportunity to put ideas together on how we can improve the health and safety environment of our workplace. We do not have to have safety problems in order to improve safety, any more than we have to be ill to become fitter.

> *An improvement in working conditions does not have to cost much or be very complicated. Simply moving a filing cabinet to a more convenient location can often represent a quantum leap towards working safely.*

Make a note here of any personal objectives that you may have.

Notes

REASONS TO LEARN

In the various studies that have been undertaken on how and why people learn and why some people learn more quickly than others, it has been discovered that motivation plays a significant role in our ability to learn.

When answering both the questions set below, please consider not only your own personal situation but those of the company, the organization, your work colleagues and, possibly, your customers.

1. *What **difficulties** or **disadvantages** derive when managers and supervisors are not aware (or are not aware as they should be) of good safety methods and practices?*

2. *What **benefits** or **advantages** derive when they are more health and safety-conscious and skilled?*

Consequences of poor safety: Consider not only business costs, but costs to you personally. These may include lost overtime and bonuses, cost of medical prescriptions, missed work opportunities and disruptions to your social life and hobbies.

> PLEASE COMPLETE
> BEFORE CONTINUING

OPINION POLL

Consider the following statements, first marking each with your level of agreement, and then making some supporting comments regarding these views.

> 5 = Strongly agree; 4 = Agree; 3 = Neither agree nor disagree; 2 = Disagree; 1 = Strongly disagree.

> 1. Every accident or injury can be prevented or avoided.

Circle one response: 5 4 3 2 1

Comments:

> 2. Every accident or work-related injury or discomfort is caused by human error in some way.

Circle one response: 5 4 3 2 1

Comments:

> 3. You cannot motivate people to be safer; you can only enforce rules and penalties.

Circle one response: 5 4 3 2 1

Comments:

> 4. Left to their own devices, people and organizations will take unnecessary risks and cut corners.

Circle one response: 5 4 3 2 1

Comments:

> **PLEASE COMPLETE BEFORE CONTINUING**

Opinion Poll: Review

> 1. Every accident or injury can be prevented or avoided.

This is largely held to be true. Research shows that nearly all accidents are a result of a cause and effect relationship. If you identify the causes, you can change the effects.

> 2. Every accident or work-related injury or discomfort is caused by human error in some way.

As a computer programmer once remarked, 'There is no such thing as "computer error", only incorrect user input'. 'Accidents' are caused by people and their behaviours, not by machines, chemicals or inanimate objects.

> 3. You cannot motivate people to be safer; you can only enforce rules and penalties.

Hopefully, people will work safely and consider their own welfare and that of others without legal or management interference, although statistics do not prove this to be the case. In countries without enforced legislation, people are made to endure terrible work environments with little or no regard for safety. Consider how many of us wear a seat belt today compared with the number who did so before it became law.

> 4. Left to their own devices people and organizations will take unnecessary risks and cut corners.

Accident investigators and HSE inspectors have thousands of examples which prove this statement to be true. You cannot have a quality company that does not consider the health and safety of its staff and customers as the highest priority.

Chapter 2
Understanding Safety Legislation and Practice

This chapter examines the current regulations and standards of safe work practice. This includes:

- understanding the Health and Safety legislation that affects you
- knowing your legal responsibilities as a manager and employee
- accident statistics
- identifying common workplace hazards and risks.

Before starting this chapter, please take a few moments to make a note of any ideas or actions in the learning diary and log in Chapter 1.

The biggest risk is not taking safety seriously.

WHAT THE LAW REQUIRES: PUTTING SAFETY FIRST

What the law requires: making safety a priority

Under the Health and Safety at Work Act (HASAWA) you must ensure the health and safety of yourself and others who may be affected by what you do or fail to do. This means making health and safety a priority.

This includes people who work for you (including casual workers, part-timers, trainees and subcontractors), use workplaces which you provide, are allowed to use your equipment, visit your premises, may be affected by your work (for example, your neighbours or the public), use products that you make, supply or import, or use your professional services (for example, if you are a designer).

The Act applies to all work activities and premises and everyone at work has responsibilities under it, including the self-employed.

New regulations have replaced and updated much of the old law on Health and Safety, but there are specific laws applying to certain premises, such as the Factories Act 1961 and the Offices, Shops and Railway Premises Act 1963.

Consider these points:

- The same mistakes that cause injury and cost lives can also damage property, equipment, delay production and inconvenience customers.

- There is no such thing as a 'minimal acceptable level' – **all** accidents must be controlled and prevented.

- A quality company is a **safe** company.

Make a note of any important points arising from this section.

Notes

_____

_____

_____

_____

_____

_____

HEALTH AND SAFETY LEGISLATION

Common law requires that an employer must take reasonable care to protect employees from risk of foreseeable injury, disease or death at work. In the nineteenth and early twentieth centuries employers argued with reasonable success against this duty. It was not until 1938 that the House of Lords identified, in general terms, the duties of employers at common law.

Whilst it is obviously good common sense to work safely, minimizing the chances of accidents, it is also a point of law. There are two main kinds of Health and Safety law. Some is very specific about what you must do, and some is much more general, applying to almost every business. In this short summary we will be looking at the key legislation that affects your work, and translating it into practical measures by which we must all abide.

Never underestimate the consequences of breaking Health and Safety legislation. We can all remember terrible accidents such as the Zeebrugge disaster or the Clapham rail accident. However, there are many hundreds of thousands of accidents each year which do not make the headlines but still ruin lives.

Legally, accidents like these can cost companies and individuals thousands and sometimes hundreds of thousands of pounds.

> ***Example:*** *A meat processing plant was burnt to the ground by a fault caused during work on a piece of machinery. The fire alarm and fire sprinklers failed to operate properly – both legal requirements. Fortunately, nobody was badly hurt.*
>
> *Any fines imposed by the courts would be insignificant compared to the loss of business, customer contracts, and the effect on 100 employees who lost their jobs for a year while the factory was rebuilt.*

Some risks are very obvious; others less so. The purpose of legislation is that all possible steps are taken to eliminate hazards, reduce risks, and inform and implement safe systems of work.

25% of all fatal accidents, and many more serious injuries, are caused because safe systems of work are not provided for, or even ignored.

Health and Safety law – what does it mean to you?

Safety regulations or legislation create very real obligations, not only on companies and organizations, but also on the directors, managers and individual employees. Whilst some are over 20 years old, many are much more recent than that and it is important that we are fully aware of the consequences and the requirements of each set of regulations. As the law says, ignorance is no defence.

In the event of an accident, Health and Safety laws are interpreted so that the company or the organization, its managers and directors, have to prove that it was not at fault. In other words, it is assumed that the organization has not met its Health and Safety obligations, unless it can prove otherwise. Therefore, failure to comply in a way that demonstrates and proves legislation has been adhered to can easily lead to prosecution, resulting in fines and, in some cases, even imprisonment.

Some legislation that may affect you:

- Health and Safety at Work Act 1974
- Electricity at Work Regulations 1989
- COSHH Regulations 1994
- Manual Handling Operations Regulations 1992
- Noise at Work Regulations 1989
- Fire Precautions Act 1971
- Display Screen Equipment Regulations 1992
- Workplace (Health, Safety and Welfare) Regulations 1992
- Management of Health and Safety at Work Regulations 1992
- Provision and Use of Work Equipment Regulations 1992
- Safety Signs and Signals Regulations 1996
- Consultation with Employees Regulations 1996

In law, ignorance is no defence

Make a note of any points from this section that concern you.

Notes

HEALTH AND SAFETY AT WORK ACT 1974

This is the main law that covers **everyone** at work and **all work premises**.

It simply means making sure that people work safely, are safe and that their welfare is not put at risk.

Enacted in 1974, this piece of legislation was brought in to replace and update much of the old Health and Safety law which was contained in the Factories Act 1961 and the Offices, Shops and Railway Premises Act 1963. These two laws were rapidly becoming out-of-date with the advent of modern working practices, technology and equipment.

Under the Health and Safety at Work Act (HASAWA) a company has to ensure the health and safety of all its employees. Individuals have to ensure the health and safety of themselves and others around them who may be affected by what they do, or fail to do. This includes contractors, as well as customers or, indeed, anybody who may come into contact with the organization.

> *Safety studies have highlighted that small firms (those with less than 50 employees) have poorer accident records than large organizations. This is made even worse by the large number of accidents that go unreported by small companies. Ignorance, poor standards and contempt for basic safety standards have been highlighted as key contributing factors.*

The Act applies to all work activities and premises and everyone at work has responsibilities under it – including the self-employed. Here are some key points raised by the Act.

1. Safeguards

Employers are required to implement reasonable safeguards to ensure safe working practice at all times. This means taking every practical step to remove hazards and reduce or eliminate risks. The law interprets this as taking every possible precaution, and cost is not considered as an excuse for failure to do this.

2. Written policy

All organizations employing five or more people must have a written and up-to-date health and safety policy. In addition, they must carry out written risk assessments as part of the implementation of their safety policy and also display a current certificate as required by the Employer's (Compulsory Insurance) Liability Act 1969.

3. Training and information

Following on from this, all staff must be fully trained, equipped and informed of not only the company's safety policy and procedures but also of the skills and knowledge necessary to carry out their normal work duties. This, of course, means displaying Health and Safety regulations and safety signs, as well as formally training and directing staff on all aspects of health, safety, hazards and risks.

4. Reasonable care

The legislation does not just cover employers and organizations; there are definite requirements placed on employees. All employees must take reasonable care not only to protect themselves, but also their colleagues. They are also required to comply with all health and safety policy regulations and procedures in full, and to cooperate fully with Health and Safety representatives and officers in their job of implementing the Health and Safety policies. Failure to do so is in breach of the Act.

HSE Inspectors can visit without notice and have right of entry. They have the power to stop work, close premises and even prosecute.

5. Safe systems of work

Employers must also ensure what are called 'safe systems of work', which means creating an environment that is conducive to health and safety. This can be as basic as making sure that buildings are in good repair, that proper heat and ventilation is provided, and that the workplace is clean and hygienic. However, it may also mean having clear procedures and checklists to make sure that safety is implemented. In some cases, a permit to work may be required to carry out certain jobs. A 'safe system of work' should also document what to do in the event of accidents and emergencies.

Make a note of any points from this section that concern you.

Notes

THERE IS NO SUCH THING AS AN ACCIDENT

Workplace accidents are caused by people. More accurately, they are caused by the things they do or don't do.

Equipment and machinery will sometimes fail, and incidents may occur which cause accidents, but they are nearly always traceable to some degree of human error, negligence or ignorance.

Workplace accidents happen to ordinary people. Although we may like to think that accidents only happen to other people or that we are somehow cleverer, better, luckier or more organized than other people, somebody in your team or department could be seriously injured or even killed. In the event of an accident caused by negligence, you yourself could also face legal or even criminal prosecution.

The reality is that an accident can happen to any one of us at any time. It could happen to you or me, it could happen today or tomorrow, next month or next year. Only by managing correct safety procedures can standards be achieved and maintained.

> ***Study assignment:***
>
> *Look around your normal workplace, and the room in which you are working now and identify as many potential hazards as you can. Find at least five.*
>
> 1.
> 2.
> 3.
> 4.
> 5.

The dictionary defines an accident as 'an unforeseen event' or 'a misfortune or mishap, especially causing injury or death'. However, the safety statistics tell us that we can predict accidents, because we know what the causes are and, if we see the causes occurring we can be sure that there is an accident waiting to happen somewhere along the line.

Accidents do not happen by themselves; they are caused by people like you or me not taking safety seriously.

How aware are you of safety in your day-to-day work as a manager?

Notes

ACCIDENT STATISTICS

How many people do you think suffer disabling injuries every year whilst at work?

a) 10 000 or less   b) 100 000   c) 1 000 000 or more

How many people do you think are killed at work every year?

a) 10 or less   b) 100   c) 500 or more

How many working days do you think are lost every year in the UK because of accidents, sickness or injury?

a) 1 million   b) 5 million   c) 10 million or more

What do you think are the three most common accidents at work?

1.

2.

3.

PLEASE COMPLETE BEFORE CONTINUING

ACCIDENT STATISTICS

For an industrialized nation like ours, with a reasonably good record of health and safety awareness, the statistics are quite surprising. Every year there are over 30 million working days lost because of work-related accidents, sickness or injury.

A casual attitude can often result in a casualty!

There are literally hundreds of thousands of workplace accidents every year, and, at any one time, several million people are suffering ill-health, either caused or made worse by working conditions. Furthermore, on average, every working day there are at least two fatal injuries in the workplace. This means that tonight somebody, somewhere, will not be going home.

Accidents can happen to any of us at any time. They are not a rarity. Fortunately, they are also not that common, but we do need to make sure that we do not become another statistic.

Workplace accidents are more common than you might think.

Here are some more painful statistics illustrating the most common types of accident and their associated causes. They were produced by the Health and Safety Commission and deal with the most commonly occurring accidents that have been reported. (There are, of course, many more – possibly a much larger number – that go unreported.)

1.	**Slips, trips or falls**	35%
2.	**Falls from height**	21%
3.	**Injuries from moving, falling or flying objects**	12%

For the self-employed

1.	**Falls from height**	45%
2.	**Slips, trips and falls (at the same level)**	15%
3.	**Injuries from moving, falling or flying objects**	14%

These statistics are for the most recent period available. However, the HSC comment that 'whilst most other accidents stayed relatively unchanged, slip, trip or fall accidents have increased from 26% to 35% for employees for the period 1986 to 1996'. These figures do not take into account illness or sickness that may be caused by repetitive strain or cumulative injuries.

So, in summary, there are 1.6 million accidents at work each year; 2.2 million people are currently suffering ill-health caused, or made worse, by work conditions; 30 million working days per year are lost; and every year about 500 people are killed at work and several thousand more are permanently disabled through work-related accidents or injury.

Ignorance and laziness will nearly always cause an accident.

The most common workplace accidents

- **Straining the body**
  – twisting, reaching or stretching

- **Moving or falling objects**
  – most common damage to head, fingers, feet and eyes

- **Slips, trips and falls**
  – from minor grazes to a broken neck

- **Getting caught in a machine**
  – belts, pulleys, slicers, grinders and so on

- **Hazardous chemicals**
  – breathing or direct exposure

- **Hearing loss or damage**
  – loud noise can destroy or damage hearing

- **Electric shock**
  – almost **any** electric tool or appliance can kill

- **Eye damage**
  – flying objects, splashing liquids, intense heat or light

A moment's carelessness – a lifetime's regret.

Find what are the most common accidents, sickness or injury in your workplace. Your Safety Officer or Personnel Manager should be able to provide this information.

Notes

WORKPLACE SAFETY AWARENESS

Many people, when first discussing the topic of workplace safety, ask the question 'But our workplaces are safe, aren't they?'. The answer, of course, is that they can be, but only if we make safety a priority. Your workplace is as safe, or as dangerous and hazardous, as you choose to make it. However, workplace 'accidents' may not be accidents at all.

Many workplace accidents are cumulative and take years to take effect. For instance, RSI, or repetitive strain injury, can be the result of years of neglect or poor practice in doing repetitive tasks – often involving very small repetitive movements as in using a keyboard or operating machinery – in an intense or prolonged fashion. The effects of these injuries are sometimes very far-reaching.

Good housekeeping

Much of good safety is really just good common sense, and as we walk around our workplaces on a daily basis we must keep in mind what could generally be summarized as good housekeeping. In 1992 a new series of regulations, the Workplace (Health, Safety and Welfare) Regulations 1992, were introduced to update the now ageing Offices, Shops and Railway Premises Act 1963 and the Factories Act 1961. These new Regulations take account of new working conditions, new working styles and new technology in the work environment. However, the basic fundamental principles remain the same – think safety; act safety; be safe.

Here are seven keys to good safety awareness.

1. *Walk areas*

Walk areas must be kept clear and tidy. This does not just apply to emergency exits, fire exits or areas through which customers may be walking. **All** work areas are kept clear and free of debris, rubbish boxes and other obstructions.

2. *Drawers*

Leaving drawers open can cause a very annoying and pointless accident – particularly as they are so simple to close!

3. *Chemicals*

Chemicals must be stored and labelled correctly, whether these are correcting fluid, floor cleaners, polishes, abrasives or specialized chemicals used in our work. Failure to do so is breaking the law.

> *Being safe simply means knowing what to do (and what NOT to do) and then DOING what you know.*

4. Ventilation and heating

The organization for which you work must provide a reasonable working temperature in all workrooms, local heating or cooling systems where a comfortable temperature cannot be maintained and good ventilation. Draughts and heating giving off dangerous or offensive levels of fumes must be avoided. In addition, workrooms should be spacious enough to work comfortably in, and arrangements should be made to protect non-smokers from discomfort caused by tobacco smoke.

5. First aid

Every workplace environment should have first aid equipment, a trained first-aider and detailed procedures to follow in the event of personal accident or injury. There should also be an accident book and a list of all uses of first aid equipment.

6. Noise and hygiene

These must be controlled. Clean and well ventilated toilets with wash basins, hot and cold running water and drinking water must be provided. Noise should either be eliminated or kept to an absolute minimum and, where noisy environments cannot be eliminated, protective equipment must be issued and worn.

Think Safety!

- **But we work safely, don't we?**
- **... only if we make safety a priority.**
- **Many 'accidents' are cumulative and take years to take effect.**
- **A moment's carelessness – a lifetime of regret!**

---

7. Safety signs

These must be provided at all appropriate points and for all valid reasons. They must be in good condition and clearly displayed. People should also be aware of their meaning, either through explanation or training. Whilst many are very self-explanatory, others are slightly more complicated and you may need to understand what these mean.

> **Safety signs:**
>
> *A safety sign with a blue background and white writing signifies a mandatory – must do – instruction.*
>
> *A safety sign with a yellow background and black writing signifies a warning – care and caution – instruction.*
>
> *A safety sign with a red background and white writing signifies a fire equipment instruction.*
>
> *A safety sign with a green background and white writing signifies safe conditions – for example, fire escapes, exits, first aid box and so on.*
>
> *A safety sign with a diamond shape and either a red, blue, yellow, white or green background and black writing signifies that a package or load contains hazardous substances.*

Make a note of any points from this section that concern you.

Notes

IDENTIFYING HAZARDS AND RISKS

Understanding hazards and risks

- A hazard is anything that has the potential to cause harm (for example, chemicals, electricity, working from ladders and so on).

- A risk is the likelihood (great or small) of harm actually being done.

As an example, consider a can of solvent on a shelf. There is a hazard if the solvent is toxic or flammable, but very little risk. The risk increases when it is taken down from the shelf and poured into a bucket. Harmful vapour is given off and there is a danger of spillage. The situation is made much worse if a mop is then used to spread it over the floor for cleaning. The chance of harm – that is, the risk – is then high.

You will see the term 'risk assessment' used in regulations and guidance. Do not be put off by this phrase – it's all about doing things described in this chapter.

Who might be at risk?

- workers – including those off-site

- visitors to your premises – for example, cleaners, contractors

- the public – for example, when calling in to buy your products.

Look for the hazards – walk around your workplace. Imagine what might go wrong at each stage of each task being carried out. Here are some typical activities which carry a risk of accidents:

- receipt of raw materials – for example, when lifting or carrying

- stacking and storage – for example, falling objects and materials, exposure to toxic substances

- movement of people and materials – for example, falls, collisions, processing of raw materials, exposure to chemicals, maintenance of buildings, such as roof work and gutter cleaning.

- maintenance of plant and machinery – for example, lifting tackle, equipment using electricity

- distribution of finished goods – for example, obstructing the movement of vehicles dealing with emergencies, spillages, fires.

In many businesses, most accidents are caused by a few key activities. Ignore the trivial and concentrate on those that could cause serious harm. But don't just look at the obvious ones – operations such as roof work, maintenance and transport movements (including forklift trucks) cause far more deaths and serious injuries each year than many mainstream activities.

Assess the risk

First, consider any accidents which you may have suffered. What happened? What was the hazard? How high was the risk? What was its nature? How could it have been minimized?

Then, consider the tasks being carried out in your own workplace:

- What is the worst result of an accident? Is it a broken finger, someone suffering permanent lung damage or death?

- How likely is it to happen? How often is the job done? How close do people get to the hazard? How likely is it that something can go wrong?

- How many people could be hurt if things did go wrong? Could this include people who don't work for you?

Don't forget non-production tasks, off-site activities and work done outside normal working hours. You should know what your main risks are.

> *Are the main risks under control?*
>
> *You now need to see if you are taking the right precautions. You may already be doing enough, but how can you be certain?*
>
> *Look at the work, talk to people and check records. Find out what actually goes on, and not what you think goes on.*

Make improvements

If you find that more needs to be done, ask yourself if you can eliminate the hazard by doing the job in a different way – for example, by using a different, safer chemical or buying materials already cut to size instead of doing it yourself. If you cannot, you should consider controlling the hazard in some other way.

- Deal with the hazards that carry most risk first.

- Set realistic dates for each of the improvements needed.

- Don't try to do everything at once.

- Remember to agree precautions with the workforce, working together to solve problems.

- Don't forget that new training and information could be needed.

- Check that precautions remain in place.

If you find that you have quite a lot to do, consider preparing an action plan, stating what you will do and when.

Remember that situations and circumstances change – new materials come in, machines wear out and break down and need regular maintenance, rules are broken and people don't always do as they've been told.

The only way to find out about such changes as these is to check. Don't wait until something goes wrong but, equally, don't try to check everything at once. Deal with a few key issues at a time, starting with the main hazards which you identified earlier. Doing this also lets people know that checks will be made and that you are interested in what is happening day-to-day in the workplace – not just when things have gone wrong.

Don't forget maintenance

Be guided by manufacturers' recommendations when working out your own maintenance schedules for such items as vehicles, fork-lift trucks, ventilation plant, ladders, portable electrical equipment, protective clothing and equipment and machine guards.

Remember, checks are no substitute for maintenance.

Safety inspectors and the law

Health and Safety laws which apply to your business are enforced by inspectors either from the Health and Safety Executive (HSE) or from your local council. Their job is to see how well you are dealing with your workplace hazards, especially the more serious ones which could lead to injuries or ill-health. They may wish to investigate an accident or a complaint.

Inspectors do visit workplaces without notice but you are entitled to see their identification before letting them in.

Don't forget that they are there to give help and advice, particularly to smaller firms which may not be well informed. When they do find problems they will aim to deal with you in a reasonable and fair way. If you are not satisfied with the way in which you have been treated, take the matter up with the inspector's manager, whose name will be given on all letters from the HSE. Your complaint will certainly be investigated, and you will be told what must be done to correct the situation if a fault is found.

Inspectors do have wide powers which include the right of entry to your premises, the right to talk to employees and safety representatives and the right to take photographs and samples. They are entitled to your cooperation and answers to questions.

If there is a problem, they have the right to issue a notice requiring improvements to be made, or (where a risk of a serious personal injury exists) one which stops a process or the use of dangerous equipment. If you receive an improvement or prohibition notice you have the right to appeal to an industrial tribunal.

Inspectors do have the power to prosecute a business or, under certain circumstances, an individual for breaking Health and Safety law, but they will take your attitude and safety record into account.

How often do you need to check?

Some important items may have to be checked daily, while others can safely be left for much longer. This is a matter for you to decide, except in cases where the law requires some inspections or examinations to be carried out by specially appointed people.

Here are some key examples:

- Ventilation systems must be examined and tested every 14 months.
- Power press guards must be inspected at each shift.
- Scaffolds must be inspected weekly.
- Rescue equipment must be examined monthly.
- Other periodic tests and examinations must be carried out by a **'competent person'**. For example, an electrician, trained to the formal and required procedure, should carry out tests on electrical equipment and installations.

Who is a competent person?

A competent person is someone who has the necessary technical expertise, training and experience to carry out the examination or test. This could be an outside organization such as an insurance company or other inspecting organization, a self-employed person or one of your own staff who is capable of doing the task and has been given the task by the company.

Make a note of any points from this section that concern you.

Notes

HAZARDS AND RISKS

Take time to reflect on the question below, making some notes in the space provided.

List ten hazards that exist in your workplace, and rate their risk (chance of happening) as either low, medium or high. (Please give examples and be specific.)

1.

2.

3.

4.

5.

6.

7.

8.

9.

10.

> **PLEASE COMPLETE BEFORE CONTINUING**

Chapter 3
A Step-by-Step Guide

This chapter deals with implementing safety and gives four steps to managing safely.

Before starting this chapter, please take a few moments to make a note of any ideas or actions in the learning diary and log in Chapter 1.

MANAGEMENT OBLIGATIONS

As a manager, director or supervisor, you must ensure that you carry out the following tasks:

- Provide a written, up-to-date health and safety policy if you employ five or more people.

- Carry out a risk assessment (and if you employ five or more people, record the main findings and your arrangements for health and safety).

- Notify your occupation of premises to your local Health and Safety Inspector if you are a commercial or industrial business.

- Display a current certificate as required by the Employers' Liability (Compulsory Insurance) Act 1969 if you employ anyone.

- Display the Health and Safety Law poster for employees or give out the leaflet.

- Notify certain types of injury, occupational disease and event.

- Consult any appointed union safety representatives on certain issues, such as any changes which might affect health and safety and any information and training which has to be provided.

Make a note here of any of the above points that give you cause for concern.

Notes

_____

_____

_____

_____

_____

FOUR STEPS TO MANAGING SAFELY

Good health and safety policy only works with the commitment and cooperation of people, and that starts with the example and leadership of managers and supervisors.

As a manager you have a responsibility to your company, your staff, shareholders and customers to manage **safely**. Failure to do so carries consequences and penalties, as well as legal fines and even imprisonment. Poor health and safety can result in lower productivity, increased sickness and absenteeism and costly breakages and mistakes. Whilst it might seem initially a little daunting, implementing health and safety is relatively straightforward. It can be summarized in these four simple steps.

- Know and understand your safety policy and procedures.
- Plan ways to reduce risk and remove hazards.
- Organize people and resources to create a safe working environment and safe systems of work.
- Measure your safety record by statistics and discussion with your staff. Investigate accidents fully.

Make a note here of three actions that you have recently taken to actively manage and implement safety standards.

Notes

_____

_____

_____

_____

SET YOUR POLICY

Your policy must be a written statement and should make reference to the above key elements listed below, together with specific preventive measures and individual responsibilities.

The same types of mistake that cause injuries and illness can also lead to property damage and interrupt production, so you must aim to control **all** accidental loss. Identifying hazards and assessing risks, deciding what precautions are needed, putting them in place and checking that they are used protects people, improves quality and safeguards plant and production.

Your health and safety policy should influence all your activities, including the selection of staff, equipment and materials, the way work is carried out and how you design and provide goods and services. A written statement of your policy and the organization and arrangements for implementing and monitoring shows your staff, and anyone else, that hazards have been identified and that risks have been assessed, eliminated or controlled.

The key elements are:

- identifying hazards (something with potential to cause harm)
- reducing risks (the chance of that harm happening)
- deciding what precautions can be taken
- putting procedures and controls in place.

Make a note of any points from this section that concern you.

Notes

_____

_____

_____

_____

SETTING SAFETY POLICY

Take a moment to answer the following key questions:

1. Do you have a clear policy for health and safety; is it written down?
2. Does it specify who is responsible, and the arrangements for identifying hazards, assessing risks and controlling them?
3. Do your staff know about the policy and understand it: are they involved in making it work?
4. Is it up-to-date?
5. Does it prevent injuries, reduce losses and really affect the way you work?

Make some notes here in response to the above.

Notes

_____

_____

_____

_____

_____

_____

_____

_____

PLAN AND SET SAFETY STANDARDS

Planning is the key to ensuring that your health and safety efforts really work. Planning for health and safety involves setting objectives, identifying hazards, assessing risks, implementing standards of performance and developing a positive culture. It is often useful to record your plans in writing.

Your planning should provide for:

- identifying hazards and assessing risks, and deciding how they can be eliminated or controlled

- complying with the Health and Safety laws

- agreeing health and safety targets with managers and supervisors

- a purchasing and supply policy which takes health and safety into account

- careful design of tasks, processes, equipment, products and so on

- safe systems of work

- procedures to deal with serious and imminent danger

- cooperation with neighbours and/or subcontractors

- setting standards against which performance can be measured.

Plan safety by...

- **setting objectives**
- **identifying hazards**
- **assessing risks**
- **implementing standards**
- **knowing the legislation**

Standards help build a positive culture and control risks. They should identify who does what, when and with what result, and apply to:

- premises, workplace and environmental controls
- plant and substances, purchase, supply, transport, storage and use
- procedures, design of jobs and the way work is done
- products and services, design, delivery, transport and storage.

Make a note of any points from this section that will help improve your own health and safety planning.

Notes

KEY POINTS ABOUT STANDARDS

There is no such thing as an accident.

The key elements of safety planning are:

- **setting objectives**
- **identifying hazards**
- **assessing risks**
- **implementing standards and targets**
- **developing a 'Safety Culture'**.

You need to have a good working knowledge of key legislation, as it relates to your workplace. This can be achieved by consulting the relevant professional bodies (the Health and Safety Executive and ROSPA, for example), attending training programmes and keeping up-to-date with safety practices and legislation.

Safety standards

Safety standards should be:

- **measurable** – who, when, how much, what results and so on
- **achievable** – based on incremental improvements
- **realistic** – people must believe them to be possible.

All plans – no matter how good – must also be sufficiently flexible to allow for the unexpected. Statements such as 'staff must be trained' are difficult to measure if you don't know exactly what 'trained' means and who is to carry out the training. 'All machines will be guarded' is difficult to achieve if there is no measure of the adequacy of the guarding. Many industry-based standards already exist and you can adopt them where applicable.

In other cases, you will have to take advice and set your own standards, preferably referring to numbers, quantities and levels which are seen to be realistic and can be checked. For example:

- maintaining workshop temperatures within a specific range

- specifying acceptable levels of waste, effluent or emissions
- specifying methods and frequency for checking guards on machines
- creating ergonomic design criteria for tasks and work stations
- specifying levels of training
- agreeing to consult staff or their representatives at specified intervals
- monitoring performance in particular ways at specified times.

Make a note of any points from this section that concern you.

Notes

_____

_____

_____

_____

_____

_____

_____

PLEASE COMPLETE BEFORE CONTINUING

IMPLEMENTING STANDARDS

Translate your standards into:

- attitudes and behaviours towards safety
- skills and expertise
- knowledge and understanding required.

This means a combination of

- training
- communicating
- organizing
- controlling
- coaching
- enforcing.

What are the standards and attitudes towards safety amongst your staff?

Notes

_____

_____

_____

_____

_____

SAFETY PLANNING AND SETTING STANDARDS

> ### *Safety planning: key questions*
>
> + What is your plan towards health and safety?
> + How do you consider safety issues?
> + Have you identified hazards and assessed risks?
> + Have you set safety targets and standards?
> + Do you have contingency plans for major disasters?

Take a moment to answer the following key questions:

1. Do you have a health and safety plan?
2. Is health and safety always considered before any new work is started?
3. Have you identified hazards and assessed risks to your own staff and the public, and set standards for premises, plant, substances, procedures, people and products?
4. Do you have a plan to deal with serious or imminent danger – for example, fires, process deviations and suchlike?
5. Are the standards implemented and risks effectively controlled?

Make some notes here in response to the above.

Notes

_____

_____

_____

_____

ORGANIZING YOUR STAFF

Take a moment to answer the following key questions:

1. Have you allocated responsibilities for health and safety to specific people?
2. Do you consult and involve your staff and the safety representatives effectively?
3. Do your staff have sufficient information about the risks which they run and the preventive measures?
4. Do you have the right levels of expertise? Are your people properly trained?
5. Do you need specialist advice from outside the organization and have you arranged to obtain it?

Make some notes here in response to the above.

Notes

_____

_____

_____

_____

_____

_____

PLEASE COMPLETE BEFORE CONTINUING

SAFETY AUDITING

As with finance, production or sales, you need to measure your health and safety performance to find out if you are being successful.

You need to know:

- where you are
- where you want to be
- what is the difference – and why.

Active monitoring – before things go wrong – involves regular inspection and checks to ensure that your safety standards are being implemented and management controls are working. **Reactive** monitoring – after things go wrong – involves learning from your mistakes, whether they result in injuries and illness, property damage or near-misses.

Safety auditing: measuring safety

- *What gets measured gets done!*
- *Learn from mistakes and near-misses quickly*
- *Set up accident investigation procedures*
- *Implement changes in legislation and standards*
- *Look behind the figures*

Two key components of monitoring systems

- **Active monitoring.** Are you implementing the standards which you have set yourself and are they effective?

- **Reactive monitoring.** Do you investigate injuries, illness, property damage and near-misses, identifying in each case why performance was substandard.

You need to ensure that information from active and reactive monitoring is used to identify situations that create risks, and to do something about them. Priority should be given to those areas where the risks are greatest.

Look closely at serious events and those with potential for serious harm. Both require an understanding of the immediate and the underlying causes. Investigate and record what happened – find out why. Refer the information to those people with the authority to take remedial action, including organizational and policy changes.

Key points

1. Regular safety audits are essential; what is measured is taken seriously.

2. Learn from your mistakes or near-misses.

3. Set up procedures to deal with accident investigation

4. Pay attention to changes in legislation, shifts in industry legal standards and/or interpretations.

5. When you are achieving your objectives and standards, raise them.

6. When compiling reports and statistics on health and safety-related matters, try to look behind the figures in order to understand the underlying causes, issues and trends.

Make a note on any points from this section that concern you.

Notes

_____

_____

_____

_____

Please complete before continuing

MEASURING SAFETY

Take a moment to answer the following key questions:

1. Do you know how well you perform in health and safety?
2. How do you know if you are meeting your own standards for health and safety?
3. How do you know that you are complying with the Health and Safety laws that affect your business?
4. How great are your losses, whether these are measured in terms of lost working time, sick leave, compensation payments or damage to plant, vehicles or equipment?
5. Do you have accurate records of injuries, ill-health and accidental loss?

Make some notes here in response to the above.

Notes

> **PLEASE COMPLETE BEFORE CONTINUING**

MANAGING SAFELY: CHECKLIST

Make a list of **everything** that you might consider when assessing your work environment and staff. You might find it useful to divide your list into each of these four categories.

1. Plant and equipment	2. Tasks and operations
3. The work environment	**4. The individual**

> **PLEASE COMPLETE BEFORE CONTINUING**

Self-assessment Worksheet

Please complete the following questionnaire, as honestly and accurately as you can. Rate your response to each statement or question on the following scale:

> 1 = Never; 2 = Sometimes; 3 = Usually; 4 = Often; 5 = Always.

1. I make sure that the company's health and safety policy is implemented and understood by all staff, visitors, contractors and temporary workers	1 2 3 4 5
2. I make sure that all staff are fully aware of their legal responsibilities with regard to health and safety at work, or while on company business	1 2 3 4 5
3. I make sure that any special directives or controls are properly carried out – for example, COSHH, use of PPE	1 2 3 4 5
4. I make sure that all staff are properly trained to an agreed or recognized standard	1 2 3 4 5
5. I make sure that job specifications and demands are within safety guidelines	1 2 3 4 5
6. I work with staff in resolving safety issues and providing a link between them and the Safety Officer	1 2 3 4 5
7. I, and my staff, attend appropriate meetings in which health and safety is to be discussed	1 2 3 4 5
8. I make sure that **all** accidents and injuries are reported according to procedures	1 2 3 4 5
9. I work with others in investigating accidents or injury	1 2 3 4 5
10. I make sure that all employees within my area are provided with accurate, timely and relevant information, training and instruction to enable them to work safely	1 2 3 4 5
11. I pay particular attention to new employees or those in hazardous areas or working with mechanical equipment or hazardous chemicals.	1 2 3 4 5

cont'd

12. I carry out regular safety reviews or cooperate actively with others who may be carrying out these reviews.	1 2 3 4 5
13. I implement completely any and all recommendations and actions as required by the safety officer, arising from these reviews.	1 2 3 4 5
14. I make sure that the correct safety equipment is made available to all employees who need to use it.	1 2 3 4 5
15. I make sure that the machines and equipment are used correctly and regularly serviced and maintained.	1 2 3 4 5

Total score:_____\100   or_____%

Analysis

Between 80%–100%   Excellent.
Between 60%–80%   Very good – a really high level of safety awareness.
Between 40%–60%   OK, but there is room for improvement.
Less than 40%   Don't walk under too many ladders, or cross the road on your own!

Make a note of the number of points you scored and remember to try to improve on this.

> **PLEASE COMPLETE BEFORE CONTINUING**

Dos and Don'ts: Managing Safely

How can you be sure that you are managing safely? List everything you should or should not be doing.

Dos	Don'ts

PLEASE COMPLETE BEFORE CONTINUING

SAFETY IDEAS

- Make sure that the company's health and safety policy is implemented and understood by all staff, visitors and contractors or temporary workers.

- Make sure that all staff are fully aware of their legal responsibilities with regard to health and safety at work, or while on company business.

- Make sure that any special directives or controls – for example, COSHH, PPE – are properly carried out.

- Organize a safety training course for all or some staff.

- Check all safety notices.

- Review job specifications and make sure that the demands of tasks fall within safety guidelines.

- Set up a meeting with staff to discuss safety issues, and involve the safety officer.

- Make sure that **all** accidents and injuries are reported according to procedures.

- Look back over accident or injury records.

- Carry out safety reviews, or cooperate with others who may be carrying out these reviews.

- Make sure that the correct safety equipment is made available to all employees who need to use it.

- Start a 'safety ideas' scheme.

- Tidy up the work area.

- Start measuring safety practices much more closely.

Chapter 4
Safety Costs!

This chapter looks at the costs of poor safety in the workplace and covers the following:

- reviewing the cost and consequences of poor safety

- understanding how good safety practice saves money

- the cost of safety to business in the UK.

Before starting this chapter, please take a few moments to make a note of any ideas on actions in the learning diary and log in Chapter 1.

HEALTH AND SAFETY: COST OR BENEFIT?

Ensuring safety procedures are correctly maintained saves time and money.

Department of Employment Labour Force Survey figures estimate that the overall annual cost to employers is between £4.5 and £9.5 billion.

This overall cost is then broken down into constituent parts – namely, damage/repair costs of £2.2–6.6 million, production costs of £566 million, staff administration costs of £444–993 million and insurance costs of £1.25 billion.

In most of these categories the specific figures for injury and non-injury accidents and cases of ill-health are detailed, with non-injury accidents accounting for the most significant proportion of the costs. (Note: a non-injury accident is one which involves damage to plant, equipment and property and so on.)

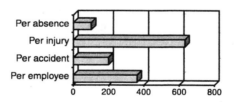

Cost of Accidents (£)

Further calculations have provided cost figures of:

- £170 to £360 per employee

- £90 to £200 per accident

- £550 to £630 per injury

- £750 to £1000 per person taking time off work or changing their job, due to ill-health

- £400 to £500 per year for each of the 1.5 million people suffering work-related ill-health.

These costs are fairly typical of accidents nowadays.

THE COSTS OF ACCIDENTS AND ILL-HEALTH

There is growing pressure from enforcement officers and insurance companies (through employer's liability insurance) for employers to manage health and safety effectively in their workplace.

Whilst there is a moral obligation to ensure that no one is intentionally harmed by their work, it is arguably the need to reduce 'bottom-line' costs, now known to be associated with occupational accidents and ill-health, which has proved most influential in persuading employers of the benefits of effective health and safety management.

The first official HSE document which addressed the subject of the costs related to health and safety at work was *The Costs of Accidents at Work* which estimated the costs of accidents over a set period of time in five different occupational situations. This work has been expanded by two more detailed HSE reports entitled *The Costs to the British Economy of Work Accidents and Work Related Ill-health* and *Self-reported Work Related Illness*.

The first of these latter two reports identifies the costs of work-related accidents and ill-health for three defined groups: individuals, employers and society. The original statistics were obtained from a health and safety trailer questionnaire to an Employment Department Labour Force Survey, which covered the previous 12-month period.

Total estimated annual costs of work-related accidents and ill-health

- Individuals — £5 billion
- Employers — Between £4.5 and £9.5 billion
- Society — Between £11 and £16 billion

To put the costs to employers into perspective, the highest employer costs are equivalent to £2 for every person in the world or between 4 and 10 per cent of the UK company gross trading profits or between £170 and £360 per person employed in the UK.

The costs to society are equivalent to between 2 and 3 per cent of total gross domestic product or a typical year's economic growth in the UK.

Total costs to employers

The £4.5–9.5 billion mentioned above can be broken down further into three constituent parts:

1. injury accidents: around £900 million
2. non-injury accidents: around £2.9–7.7 billion
3. ill-health: around £600 million.

The non-injury accidents – that is, damage to plant, equipment and property and so on – are included in the overall total as they are usually the result of the same underlying management failures as injury accidents and cases of ill-health.

UNHEALTHY STATISTICS

Moving away from the financial costs, it may be interesting to note some of the other statistics which are associated with, and form part of, the overall costs.

These other statistics include:

- 1.6 million injury accidents of which 640 000 people were absent for three or more days

- 2.2 million people suffering ill-health caused, or made worse, by work

- 27 million non-injury accidents

- 30 million working days lost:
  – 18 million through accidents
  – 11.6 million through ill-health.

There are more working days lost through accidents and ill-health than are ever lost through industrial disputes.

THE MOST COMMON ACCIDENTS AND THEIR CAUSES

The following information will illustrate the most common types of accident and their associated causes and, hopefully, indicate how employers can use their own accident and ill-health records to see where changes can be implemented.

Figures from the Health and Safety Commission's *Annual Statistics Report* show that the most commonly occurring accidents to employees reported to enforcement officers under RIDDOR are, in terms of percentage of all accidents:

1.	**slips, trips or falls (on the same level)**	**35%**
2.	**falls from height**	**21%**
3.	**injuries from moving, falling or flying objects**	**12%**

The same report reveals that the four most common accidents to self-employed workers are the same as above although their prevalence is differently distributed:

1.	**falls from height**	**45%**
2.	**slips, trips or falls (on the same level)**	**15%**
3.	**injuries from moving falling or flying objects**	**14%**

These statistics are for the most recent period available. However, the HSC comment that 'whilst most other accidents stayed relatively unchanged, slip, trip or fall accidents have increased from 26% to 35% for employees for the period 1986 to 1996'. All these figures include fatalities, major injuries and injuries resulting in three or more days' absence from work.

The most common injuries are musculo-skeletal injuries, particularly those affecting the back.

Estimates suggest that around 3.6 million working days per year are lost as a result of back injuries caused or made worse by work. The number of working days lost from falls from a height and machinery-related injuries are thought to be around 1.7 million and 0.75 million, respectively.

The HSE has produced statistics on self-reported work related illness, derived from a Labour Force Survey, which confirm that musculo-skeletal injuries are the most common injuries for both manual and non-manual occupations.

In manual occupations the next most common work-related illness reported were the long-term consequences of trauma and poisoning (which were the biggest causes of workers changing jobs), lung diseases, including asthma, and deafness. In non-manual occupations musculo-skeletal disorders were followed by stress and depression as the second most common complaint and, in offices specifically, headache and eye strain.

Although these particular figures are based on self-reported work-related illness – that is, the individuals' perception of their own illness and the relationship with their work – they are broadly in line with other findings.

Make a note of any points from this section that concern you.

Notes

_____

_____

_____

_____

_____

SAFETY IS FREE ...

... Accidents are the real cost

Clearly, work-related accidents and cases of ill-health have considerable impact on a company's finances. What is perhaps more surprising is that, in many companies, these costs are accepted as 'normal' company expenditure and are effectively 'written off', even though, in many situations, the causes of these accidents and cases of ill-health are preventable. Consider the following compensation awards:

- A worker who suffered a back injury caused by slipping while carrying out a manual handling operation was awarded £42 200.

- A nurse who suffered a back injury as a result of lifting a patient was awarded around £124 000.

- A machine operator who developed repetitive strain injury (RSI) as a result of placing components into another machine was awarded around £60 000.

- A factory worker who developed tenosynovitis as a result of repetitive production work was awarded £20 000.

- A fireman who suffered post-traumatic stress disorder after the King's Cross fire was awarded around £148 000.

- A worker involved in treating wood and who developed cancer as a result of dioxins in the spray was awarded £90 000.

Although these costs are insured, and therefore recoverable from the insurance company, the substantial awards that have been, and are being, made by the civil courts are now a very significant factor in the overall employer costs of occupational accidents and illness.

Companies with poor histories of personal injury or ill-health claims will be penalized by ever-increasing insurance premiums until such time that insurance companies will be unwilling to insure the risks and, as mentioned earlier, companies cannot legally operate without the employer's liability insurance.

Other insurances

The other two common insurance covers which employers take out are public liability insurance and product liability insurance. Both these insurance covers are voluntary and are fairly readily available. Insurance claims, excluding employer's liability, account for £355 million as a result of accidental fires and/or explosions and £150 million relating to other corporate liability claims.

The total annual cost of insurance claims made against employers is estimated to be around £1.25 billion.

Uninsurable costs

The most obvious uninsurable costs – that is, costs which have to be borne by the company and which cannot be recovered through insurance policies – are fines resulting from enforcement prosecutions and legal fees.

Fines

An amendment to the Health and Safety at Work Act increased the level of fines which can be imposed on summary conviction (magistrates' court level) to £20 000 per offence for certain offences. Fines for conviction on indictment (Crown Court level) remain unlimited. Furthermore, magistrates can now impose a prison sentence of up to six months, again for certain offences, while at the Crown Court a sentence could be up to two years, for specified offences.

Most Health and Safety prosecutions are brought under section 2 (Duty to Employees) and section 3 (Duty to Non-employees) and are heard by a Magistrates' Court. Where the offence is particularly serious, or where the defendant specifically requests it, the case may be referred to the Crown Court before a jury. Over recent years there has been a marked increase in the level of fines being imposed by the criminal courts. Examples include:

- a fine of £12 000 plus £1342 costs for failing to ensure a safe system of work, which resulted in a permanent knee injury during a manual handling operation

- a fine of £200 000 plus £15 000 costs for failing to ensure the safety of a worker who was crushed between two trains in the Channel Tunnel

- a fine of £250 000 plus £92 000 costs for failing to pack explosives safely.

Fines of this magnitude are still the exception rather than the rule and are rightly associated with the more serious offences. However, the average fines per conviction are rising rapidly now that higher fines are available to magistrates as detailed above.

With the replacement of prescriptive legislation by broader goal-setting laws, section 2 of HASAWA (Duty to Employees) has become even more significant and is increasingly used by enforcers to bring prosecutions. In order to fulfil the duty imposed by section 2, employers need to provide evidence of effective health and safety management in their undertaking. Recent court cases have resulted in the imprisonment of safety officers, the fining of directors and increasing fines for breaches of the law.

Other uninsurable costs

The other uninsurable costs tend to have a less direct financial effect and include:

- the injured person's lost production time

- the wages paid to persons who assist the injured person or who stop work out of sympathy or curiosity

- the wages paid to persons who are unable to continue their work as a result of the accident

- any overtime payments made in order to meet customer demands or scheduled production targets affected by the accident or incident

- damages, repairs or changes necessary to plant, equipment or materials

- supervision time spent in assisting in any investigations, reporting, reassigning work tasks and making necessary adjustments

- time spent on administration (processing investigations and reports, completing statutory forms)

- recruiting, training, instructing and supervising replacement staff or retraining and rehabilitating sick or injured staff and consequent lower initial productivity of new staff or rehabilitated staff

- provision of first aid or other medical facilities

- damage to reputation and customer confidence resulting in loss of market

- psychological effects on other employees.

Make a note on any points from this section that concern you.

Notes

THE ACCIDENT ICEBERG

The HSE report, *The Costs to the British Economy of Work Accidents and Work Related Ill-health*, estimates values for damage/repair costs, staff/administration costs and productivity-related costs.

The total damage costs are estimated to be between £2.2 and £6.6 billion, of which £17–140 million relate to accident injuries and £2.2–6.5 billion relate to non-injury accidents. The difference in the figures is partially explained by the fact that most injury accidents are caused by slips, trips, falls and manual handling and do not involve plant, equipment or materials (which are generally expensive items).

The total costs relating to productivity are estimated to be £566 million which comprises injury accidents costs of £336 million (which works out at the equivalent of £17 per day) and ill-health costs of £230 million (which is equivalent to £20 per day).

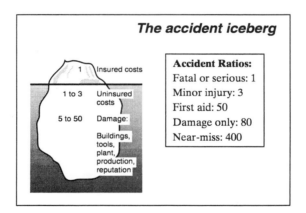

The productivity costs derive from:

- absence from work – that is:
  – maintaining output
  – sick pay
  – labour and additional administration

- actual lower productivity resulting from:
  – machine downtime
  – reorganization of staff and work procedures
  – low staff morale
  – the possible cumulative effects of ill-health.

The total costs relating to staff and administration are estimated to be between £444 and £993 million of which:

- injury accidents account for between £58 and £69 million
- non-injury accidents account for between £307 million and £712 million
- ill-health accounts for between £79 million and £212 million.

The administration costs relating to injury accidents and ill-health are £54 million and £35 million respectively, which is equivalent to £3 per day of absence in both cases.

Make a note of any points from this section that concern you.

Notes

_____

_____

_____

_____

_____

Chapter 5
Creating a Safety Culture

This chapter explains how to create a safety culture within your organization. It covers:

- identifying safety attitudes
- encouraging safety skills on a day-to-day basis
- building safety into normal working procedures.

Before starting this chapter, please take a few moments to make a note of any ideas or actions in the learning diary and log in Chapter 1.

HOW TO MANAGE SAFETY

An effective health and safety policy depends on involved and committed staff. This is often referred to as 'health and safety culture'.

There are four components of a good, or positive, safety culture:

- **Competence:** recruitment, training and advisory support
- **Control:** allocating responsibilities and securing commitment
- **Cooperation:** between individuals and groups
- **Communication:** verbal, written and visible.

How do you currently manage or 'enforce' health and safety standards? Which of the above methods do you use most?

Notes

_____

_____

_____

_____

_____

_____

_____

The safety culture: a review of the key points

Competence

- Assess the skills needed to carry out all tasks safely.

- Provide the means to ensure that **all** employees, including temporary staff, are adequately instructed and trained.

- Ensure that people on especially dangerous work have the necessary training, experience and other qualities to carry the work out safely. Arrange for access to sound advice and help.

Control

- Lead by example, demonstrate your commitment and provide clear direction.

- Identify people responsible for particular health and safety jobs – especially where special expertise is called for.

- Ensure that foremen and supervisors understand their responsibilities.

- Ensure that all employees know what they must do and how they will be supervised and held accountable.

Cooperation

- Consult your staff and their representatives.

- Involve them in planning and reviewing performance, writing procedures and solving problems.

Communication

- Provide information about hazards, risks and preventive measures.

- Discuss health and safety regularly.

- Share statistics and information openly.

- Provide clear, plentiful and accurate information regarding hazards, risks and preventive measures – an informed individual is a responsible one.

THE COST OF SAFETY

We all have many reasons to learn to be safety-conscious and we all know the consequences of poor safety and some benefits of good safety. Of the two, which would you think is the greater motivation – the consequences of poor safety or the benefits of good safety?

In fact, the consequences of poor safety tend to motivate us slightly more than the benefits of good safety. For instance, imagine that you are chopping an onion at home. The thought of slicing into your finger would probably motivate you more to be careful than the benefits of having a neatly sliced onion. So let us consider the cost of poor safety.

1. Discomfort and pain

When accidents happen to us or we injure ourselves either at work, in the home or anywhere else, we generally experience discomfort and pain. Quite often this pain is extreme. Anyone who has ever slipped a disc, dislocated a shoulder, broken a leg or suffered serious cuts and bruises will appreciate the degree of pain involved. We tend to learn very quickly after any one of these experiences.

2. Work disruptions

Accidents in the workplace can cause all kinds of disruption to our normal work schedule. An accident during the day might mean that we waste hours of working time while we receive medication. The fact that we cannot carry out our normal duties may have a snowball effect, stopping other people working and disrupting all kinds of activity. Orders may be delayed, typing may not be done, schedules may be missed and appointments postponed.

With a more serious injury, we may have to take time off work. We may not be able to play our favourite sport or pursue our favourite hobby. We may have to postpone holiday plans. All these are consequences of not taking health and safety seriously enough. In many situations – for example, in offices – instead of accidents, people experience stress, bodily aches and continual tiredness; this means that they are less effective both during the working day, and at home.

3. Money

Most managers and organizations are very concerned with money, budgets and resources. Accidents, injuries and even near-misses cost money. This cost can be in terms of damage to machinery, lost wages, higher insurance, lost customers, and possibly even compensation payments made to employees or customers.

4. Lost productivity and business

Accidents and injuries can lead to less productivity, an increased workload for others, orders being missed and customers being inconvenienced – none of which is good for business.

So you can see that the consequences of poor safety are quite considerable. We have all read about the major disasters that have happened in the last few years. Whilst train crashes, fires in underground stations, sea and river disasters all grab the headlines, every day there are thousands of accidents in workplaces just like yours that cost millions of pounds and many hours of discomfort and pain.

Make a note of any costs incurred from poor safety that you have experienced or know of.

Notes

_____

_____

_____

_____

_____

_____

_____

ATTITUDE MAKES THE DIFFERENCE!

It seems that accidents do not just happen, something causes them, and that something is usually people. To be more specific, bad attitudes often cause accidents. Here are four keys to a good safety attitude.

Good safety attitudes

1. Be informed

Take time to understand and think about the hazards and risks that exist, not only in your own work environment, but also in the various tasks and jobs you and your staff do during the day.

2 Be aware of safety procedures

Every organization has safety rules and procedures, including yours. It is very important that you, as a manager or supervisor, learn the rules concerning health and safety and then keep to them. Not only is it clearly unsafe to break safety policies and procedures, you may actually be breaking the law and therefore be liable to prosecution.

3. Cooperate with safety representatives

The Health and Safety at Work Act states that it is the responsibility of employees and managers to cooperate fully with safety representatives and managers who are implementing safety policy and procedures. To put it in a more positive way, you have the most to benefit from good safety, so it makes sense to cooperate with other people around you who are working towards that goal.

4. Be alert

Accidents happen when we walk around with our blinkers on, so always be alert, awake and aware of what is going on around you and of what hazards may exist, or potentially exist, in all situations.

Attitudes that cause accidents and injury

Now that we have considered some positive attitudes that can help us prevent accidents and injuries, let us look at the more negative attitudes that can cause accidents in the first place.

These can apply to managers and supervisors as well as staff at every level. The difference is one of scale. A manager displaying any of these attitudes can put the safety and welfare of all their staff at risk

1. Overconfidence

Overconfidence means thinking that accidents cannot happen to us, that they only happen to other people and believing that we are 'too clever' or we are 'too good'.

2. Laziness

There is a saying that states 'A short-cut is only a fast route to a shortcoming'. Do not try to cut corners or take unnecessary risks. You may be risking your own life as well as somebody else's.

3. Stubbornness

Many jobs require people to wear personal protective equipment (PPE) such as hard hats, ear protection, special clothing, safety boots and so on. In fact, we all wear a particular form of protective equipment each time we make a journey by car – our seat belts.

For example, how many people used to wear a seat belt for every journey before it was made law? Well, statistics show that it was very few of us, and the reason is probably stubbornness. Even though we all understood that it is better to wear a seat belt than to fly through the windscreen in the event of an accident, it actually took an Act of Parliament to make people do so for every journey.

4. Impatience

Many accidents are caused by people trying to do a task too quickly without paying due care and attention to what is going on around them or the consequences of not carrying it out properly. Very often we end up having to do something twice because we did not do it properly the first time. As the old saying goes, 'More haste, less speed'.

5. Ignorance

Clearly, if someone is ignorant of the dangers inherent in any particular operation or task, it might be unfair of us to expect them to know how to behave safely. However, the law says that ignorance is no defence. This is the purpose of this workbook. At the end of this training you will know what hazards and risks exist in your workplace. You will know how to prevent them and you will known how to ensure safety in your own workplace.

When a Health and Safety Inspector investigates a company after an accident or injury, the company must prove that its employees were sufficiently trained and had the right knowledge and skills for the jobs and operations that they were carrying out. Ignorance is no defence when it comes to health and safety, so make sure that you really understand and know the hazards and risks that exist in your workplace.

6. Showing off

A moment's levity can lead to long-term regret. We have all occasionally done something foolish and, fortunately, probably escaped too serious a consequence. Nevertheless, always be aware that when we let our guard down we increase the risks of accidents.

7. Forgetfulness

We are all probably guilty of this from time to time. Even when we know what we should be doing, even with our normal best intentions, occasionally we forget. Safety is a full-time job, and it requires your full-time attention. Always consider safety; never forget its importance.

Make a note of any bad attitudes to safety that you have observed in your workplace and how you might improve them.

Notes

_____

_____

_____

_____

_____

_____

_____

_____

CAUSES OF ACCIDENTS AND INJURIES

Let us look at the six key causes of accidents and injuries in offices, hospitals, schools, warehouses, factories and the working world in general.

1. Carelessness and bad habits

There is nothing particularly sophisticated about this, but it is the cause of more accidents than almost all the other factors put together. Avoid carrying out tasks wrongly or carelessly. There is a saying in Health and Safety which says that 'a casual attitude produces a casualty'. Just make sure that you or your staff are not that casualty.

2. Breaking the rules

When you know what to do, make sure that you do it.

3. Not knowing the rules

The law states that it is the organization's responsibility to train everybody – including suppliers, contractors and customers that may come in contact with you – in safety rules, policies and procedures. If you do not know, it is **your** responsibility to find out and also your responsibility to encourage this attitude in your staff.

4. Faulty equipment

Whether it is a faulty chair which gives us backache, or a faulty computer which gives us an electric shock, anything mechanical, moving or electrical has enormous potential to cause us harm and needs to be treated with great respect. If any equipment becomes faulty in any way, it is your responsibility to stop it being used, isolate it and have it put right or replaced.

5. Not thinking safety all the time

Accidents do not take coffee breaks nor do they take days off. The next accident could be a year or three seconds away – we never know. We must never – any of us – stop thinking safety all the time.

6. Thinking that it cannot happen to us

Whenever you think this, you are at your most susceptible to accident, injury or ill health. Of course, the truth is that an accident can happen to any one of us, any day. We must constantly be aware of the hazards and risks in our work environment and know how we can minimize those risks and make it safer and healthier for everybody.

Be safe!

- You have the most to gain from working safely.
- You have the most to lose.
- You are responsible for your own safety and those around you.
- All accidents can be avoided.
- All hazards should be reduced or eliminated.
- Risk must be minimized at all times.
- Don't take chances if you don't want to be a loser.

Make a note of how you can reduce the risk of accidents in your workplace

Notes

_____

_____

_____

_____

_____

_____

_____

Chapter 6
Learning Review

This chapter tests your knowledge of safety issues in the workplace.

Before starting this chapter, please take a few moments to make a note of any ideas or actions in the learning diary and log in Chapter 1.

TEST YOUR KNOWLEDGE (1)

1. List three hazards with potential to cause harm in your workplace.

    1.

    2.

    3.

2. List three common causes of accidents in the workplace.

    1.

    2.

    3.

3. What is the difference between a hazard and a risk?

4. List three of the seven keys to good safety awareness in the workplace.

    1.

    2.

    3.

5. What do the following safety signs signify?

    a) a safety sign with a blue background and white writing

    b) a safety sign with a yellow background and black writing

    c) a safety sign with a red background and white writing

    d) a safety sign with a green background and white writing

    e) a safety sign with a diamond shape and either a red, blue, yellow, white or green background and black writing

> **PLEASE COMPLETE BEFORE CONTINUING**

TEST YOUR KNOWLEDGE (2)

1. What does HASAWA stand for?

2. Small firms (employing less than 50 employees) have a worse record of safety than larger firms.

   TRUE/FALSE

3. Why do you think this is?

4. How many accidents occur at work every year?

   a) 600 000

   b) 1 600 000

   c) 2 200 000

5. How many people, at any one time, are suffering ill-health, either caused or made worse by work conditions?

   a) 600 000

   b) 1 600 000

   c) 2 200 000

6. How many working days are lost each year due to health and safety-related accidents, sickness or injury?

   a) 10 million

   b) 20 million

   c) 30 million

7. What is the most common form of accident/injury?

8. How many people are killed at work every year?

   a) 200

   b) 400

   c) 600

> **PLEASE COMPLETE BEFORE CONTINUING**

TEST YOUR KNOWLEDGE (3)

1. List three things that employers are legally bound to do.

    1.

    2.

    3.

2. List three things that employees are legally bound to do.

    1.

    2.

    3.

3. A Safety Inspector has right of entry, right to interview and take samples with or without an organization's permission.

    TRUE/FALSE

4. List three pieces of recent EU legislation that may affect you.

    1.

    2.

    3.

5. List three essentials that must be provided for in the workplace.

    1.

    2.

    3.

6. What are the most common accidents or injury in your organization (you will need to research this separately)?

7. How often should safety assessments/inspections be carried out?

8. Who is qualified to carry these out?

9. If an accident takes place at work, what is the legal position of the company?

> **PLEASE COMPLETE BEFORE CONTINUING**

CASE STUDY (1)

Read the following case study and decide how you would tackle the situation if you were Mike, listing clearly the steps that you would take and how you would solve the problem. Use the question sheet on the following page.

Getting safety taken seriously

After returning from a meeting at head office, Mike was thinking very hard about how safe his branch was. At this meeting there had been a short presentation from the company's Safety Officer regarding what they should be doing in their branches. This included training, fire drills and safety inspections, which were all news to Mike.

Filled with good intentions, he decided that it would be a good idea to make a start on some of these things, but the poor reaction he received came as a complete surprise.

First, he decided to run a test fire drill. He let all the staff know that it would happen sometime during the afternoon and that they were to assemble at the front of the building. He set the fire alarm off and made sure that he was one of the first outside, trying to lead by example, as it were. It took over seven minutes for the rest of the staff to trudge reluctantly out, several saying they were on the phone at the time and wanted to finish their conversation!

After this shambles he decided to take his clipboard and walk around the offices and warehouse looking for safety hazards and ways to improve things. He did not like what he found. There were boxes stacked next to ventilation grilles, rubbish shoved into corners, fire extinguishers hidden in cupboards, cables trailing across office floors, swivel chairs being used to reach high shelves, sockets overloaded and so on. Most worrying of all, when he tried to draw people's attention to these things he was met by a wave of indifference and comments like 'Well it's always been like that!'.

He knew that he ought to do something and do it quickly, before either the Safety Officer came round or, worse still, there was an accident.

Case Study (1): Questions

1.   What would you have done?

2.   What would you do now?

> **Please complete
> before continuing**

Case Study (2)

Consider the following problem and make notes regarding your ideas and proposals to improve or resolve the situation, using the question sheet on the following page.

Your secretary, who normally has an excellent standard of accuracy and speed of working has recently been making small errors and falling behind her work schedule. You are also aware that her timekeeping is slipping.

When you tackle her about this, she mentions that, during the day, she often suffers from headaches and finds it hard to concentrate at the computer. She suggests that it might be something to do with her computer or office environment.

It is obviously important not to dismiss this out of hand, and you need to look more closely at the situation.

CASE STUDY (2): QUESTIONS

1. Describe in detail how you would go about solving this problem.

2. Would you assess her work station and office environment? If so, create a checklist that you might use.

3. What could happen if you did nothing?

> **PLEASE COMPLETE BEFORE CONTINUING**

CASE STUDY (3)

Consider the following problem and make notes regarding your ideas and proposals to improve or solve the situation.

You are an assistant manager of a high street bank branch and have recently been given responsibility for assessing and commenting on health and safety standards in your branch.

Your manager has asked you to outline your initial approach, highlighting what you consider to be the main hazards, what risks are associated and what type of training or assessments might be required.

Notes

> **PLEASE COMPLETE BEFORE CONTINUING**

Case Study (4)

Consider the following problem and make notes on how you would handle the meeting with the goal of making sure that Bob keeps to the rules.

You are the project manager for a building company. You have recently been aware that your site supervisor, Bob, has been breaching regulations. On two occasions you have visited sites to find Bob wearing a baseball cap instead of a safety helmet and trainers instead of steel-capped boots, with which he has been issued.

Bob is an otherwise excellent employee, and puts in many long hours to ensure that work is finished on time and to standard. Bob's argument is that he has never had an accident in 15 years of working on building sites, so he doesn't need to wear the equipment which he claims is too hot and uncomfortable.

You are due to see Bob later today.

Notes

_____

_____

_____

_____

_____

_____

_____

PLEASE COMPLETE BEFORE CONTINUING

Case Study (5)

Consider the following problem and make notes on your ideas and proposals. Then use the following page to outline your presentation.

The company requires that, within two weeks of joining, all new staff are given a short presentation to outline the company's health and safety policy and procedures and to highlight some key points on how to work safely.

As the safety officer is away on holiday, you have been asked to give the talk in his absence.

Notes

CASE STUDY (5): PRESENTATION OUTLINE

Outline your presentation, briefly listing the points which you consider are important to include.

PLEASE COMPLETE BEFORE CONTINUING

CASE STUDY (6)

Consider the following problem and make notes on your ideas and proposals. Then complete the exercise on the following page.

Your safety officer recently noted that you have no written guidelines or procedures concerning 'working with ladders' – something that is done frequently by your staff.

He has therefore asked you to prepare a simple list, probably a page or two in length, of safe working practices when working from heights using ladders. He assures you that this is basically common sense and should be quite easy to do.

Notes

Case Study (6): Exercise

Prepare a first draft of your safety guidelines.

Please complete before continuing

Appendix
Suggested Answers to the Knowledge Tests

Test your knowledge (1): suggested answers

1. Answers could include, but are not limited to:

    Electric equipment
    Lifting operations
    Hazardous chemicals
    Working from heights
    Compressed air equipment
    Poorly maintained or serviced tools and equipment
    Fork-lift trucks

2. Any three of the following:

    Slips, trips and falls
    Manual handling
    Working from heights – falls
    Moving (falling/flying) objects

3. A hazard is a danger; a risk is the chance of that danger turning into an accident or injury.

4. Any three of the following:

    Walk areas kept clear and tidy
    Drawers not left open
    Chemical stored and labelled correctly
    Good ventilation and heating
    Clearly displayed safety signs
    First aid provision
    Controlled noise and good hygiene

5. a) A 'must do' instruction
   b) A warning – care and caution – instruction
   c) A fire equipment instruction
   d) Safe conditions – for example, fire escapes, exits, first aid box and so on
   e) Hazardous substances warning

Test your knowledge (2): suggested answers

1. Health And Safety At Work Act

2. TRUE

3. Perhaps safety is given a lower priority

4. b) 1 600 000

5. c) 2 200 000

6. c) 30 million

7. Manual handling, followed by slips and trips

8. c) 600

Test your knowledge (3): suggested answers

1. Any three of the following:

   Provide adequate training and supervision
   Provide PPE if needed
   Create safe systems of work
   Remove, as far as possible, all hazards and risks
   Have a written safety policy

2. Any three of the following:

   Follow safety rules
   Use all care and consideration
   Wear safety equipment that is provided
   Cooperate fully with Health and Safety officers and representatives

3. TRUE

4. Any three of the following:

   Health and Safety at Work Act 1974
   Manual Handling Operations Regulations 1992
   Display Screen Equipment Regulations 1992
   Electricity at Work Regulations 1989
   COSHH Regulations 1994
   Noise at Work Regulations 1989
   Fire Precautions Act 1971
   Workplace (Health, Safety and Welfare) Regulations 1992
   Management of Health and Safety at Work Regulations 1992
   Provision and Use of Work Equipment Regulations 1992
   Safety Signs and Signals Regulations 1996
   Consultation with Employees Regulations 1996

5. Answers could include, but are not limited to:

   Heat and ventilation
   Toilet and washing facilities
   Safety notices
   Fire exits
   Fire-fighting equipment
   Trained first-aider

6. Student's own response.

7. Every 12 months, or if conditions or work routines change significantly.

8. A 'competent' person.

9. It has to prove that it took all reasonable precautions.

Case studies

The purpose of these case study exercises is for students to apply safety knowledge and awareness.

Whilst there are no 'right' answers, students should highlight legal regulations and standards that have been broken and practical ways of enforcing these. They should also show an awareness and understanding of how to implement and maintain health and safety policies in the workplace.

HEALTH AND SAFETY WORKBOOKS

The 10 workbooks in the series are:

Fire Safety	0 566 08059 1
Safety for Managers	0 566 08060 5
Personal Protective Equipment	0 566 08061 3
Safe Manual Handling	0 566 08062 1
Environmental Awareness	0 566 08063 X
Display Screen Equipment	0 566 08064 8
Hazardous Substances	0 566 08065 6
Risk Assessment	0 566 08066 4
Safety at Work	0 566 08067 2
Office Safety	0 566 08068 0

Complete sets of all 10 workbooks are available as are multiple copies of each single title. In each case, 10 titles or 10 copies (or multiples of the same) may be purchased for the price of eight.

Print or photocopy masters

A complete set of print or photocopy masters for all 10 workbooks is available with a licence for reproducing the materials for use within your organization.

Customized editions

Customized or badged editions of all 10 workbooks, tailored to the needs of your organization and the house-style of your learning resources, are also available.

For further details of complete sets, multiple copies, photocopy/print masters or customized editions please contact Richard Dowling in the Gower Customer Service Department on 01252 317700.